Kitchen Science

**SHAR LEVINE
& LESLIE JOHNSTONE**

Illustrated by Dave Garbot
Photography by Michael Hnatov

Sterling Publishing Co., Inc.
New York

This is my 37th book, I think. I've lost count. I've dedicated books to everyone I know and love and I've even dedicated several books to some folks. My friends, family, husband's law firm, and even my kid's schools have been covered. So, with L. J.'s indulgence, this book is for Jim Becker and Charlie Nurnberg, who are kind enough to publish us, support us, and always give us great editors. A special thanks to Betsy Pringle. —S. L.

For Mark, with love.—L. J.

Acknowledgments

Thanks to C2 of Denver, Colorado, for food safety advice.
Thanks to our models and their families for their patience and help.

Photography by Michael Hnatov Photography
Photomicrographs on page 49 by the authors.
Edited by Isabel Stein
Design and layout by Judy Morgan

Library of Congress Cataloging-in-Publication Data

Levine, Shar, 1953-
    Kitchen science / Shar Levine & Leslie Johnstone ; illustrated by Dave Garbot.
        p. cm.
    Includes index.
    Summary: Shows how to turn your kitchen into a laboratory and perform
all sorts of experiments with food, such as making sun tea, creating an
acid/base tester, and gathering spores from mushrooms.
    ISBN 1-4027-0332-5
    1. Science—Experiments—Juvenile literature. [1.
Science—Experiments. 2. Food—Experiments. 3. Experiments.]
I. Johnstone, Leslie. II. Garbot, Dave, ill. III. Title.
Q164 .L474 2004
507'.8—dc21

                                                                                              2003008841

10   9   8   7   6   5   4   3   2   1

Published by Sterling Publishing Co., Inc.
387 Park Avenue South, New York, NY 10016
©2003 by Shar Levine and Leslie Johnstone
Distributed in Canada by Sterling Publishing
℅ Canadian Manda Group, One Atlantic Avenue, Suite 105
Toronto, Ontario, Canada M6K 3E7
Distributed in Great Britain and Europe by Chris Lloyd at Orca Book
Services, Stanley House, Fleets Lane, Poole BH15 3AJ, England
Distributed in Australia by Capricorn Link (Australia) Pty. Ltd.
P.O. Box 704, Windsor, NSW 2756, Australia

Sterling ISBN 1-4027-0332-5

Contents

Introduction 4

Note to Parents and Teachers 5

Safety First! 6

Materials 7

Lab in a Box 8

Salted Popcorn 10

Pierced Potatoes 12

Sun Tea 14

A Penny for Your Thoughts 16

Colorful Cabbage "Soup" 18

Now You See It, Now You Don't 21

Fairy Rings 23

Self-Inflating Balloons 25

Saltwater Daffy 27

Hard as a Rock 29

Rainbow Bursts 31

Bag It 33

Milky Whey 36

Shake, Shake, Shake! 38

Brown Bagging 40

Salty Sidewalks 42

Fizzies 44

Playing with Dough 46

Sally Sells Sea Salt by the Seashore 48

Carrot Top 50

Which Way Is Up? 52

Tree in a Jar 54

Popcorn on the Cob 56

Racing Colors 58

Taking Stalk 60

Rusting Apples 62

Genie in a Bottle 64

How Now, Green Cow? 66

Designer Foods 68

Name That Food 70

Birds of a Feather 72

The Three Bears 74

Where's Dinner? 76

Glossary 78

Index 80

Introduction

Inside your home is a humble science lab, yearning to be discovered. Where do you think you could find such an exciting place, right under your nose?

Wander down to the kitchen. Take a look around. You are in your new laboratory. You may not find a mad scientist working there (although a frantic mother preparing a meal may sometimes resemble Frankenstein's monster), but the kitchen is a perfect place to discover the wonders of simple science.

Despite the title of this book, you don't really need a whole kitchen to do these exciting science activities. You will need an adult helper, a flat work surface, and some basic kitchen materials. For some activities you need a freezer or a refrigerator. You won't need anything dangerous to do these experiments. Your mom, dad, or another helpful adult should read the experiment with you and help you with all the steps. They will have as much fun as you. Make sure you don't try to use the oven, stove, microwave, toaster, or any other kitchen appliance or sharp object for the activities in this book.

You probably don't have a spare lab coat hanging around the house, but an adult's old shirt works like a charm. Use a shirt that is never going to be worn for everyday use again. Have the sleeves trimmed off or rolled up to fit you, and voila! You look just like a scientist. If the shirt is dragging on the ground, your parents may also have to cut off the bottom of the shirt so you don't trip.

Get ready to see what you can do with common cooking ingredients. You are about to become a scientist.

• •

Fun Kitchen Science Fact: Egyptian Ice

For thousands of years, people have known that keeping foods cool helps keep them from spoiling. If you lived in Egypt around 2500 years ago, you would have put out large clay pots filled with water on cold nights. The water would freeze and you would have been able to use this ice to keep your food cold.

• •

NOTE TO PARENTS AND TEACHERS

Children learn best when they are actively involved. This book is designed to help young children begin to discover some very basic scientific principles. Adult supervision is required, as young children may not understand all the instructions in the book and may need help with some things—for example, pouring things from large bottles or bowls and cutting things. Make certain that you have gone over all the steps and that your child knows how to do the activity. It is important to tell your budding scientist that rummaging through the kitchen cabinets to find "chemicals" is not a safe thing to do. Children may be fascinated with various kitchen appliances such as the microwave or stove top, so it is critical that you keep a close eye on young scientists, who may want to explore these items.

Unless the instructions specifically say it is alright to do so, do not taste, drink, or eat the experiments. It's very important to review the safety rules (pages 6 to 7) with your children and to supervise them closely to ensure that these rules are not broken. Don't worry about the inevitable mess on the floor and counters, but try to clean up along the way to avoid slipping. That's

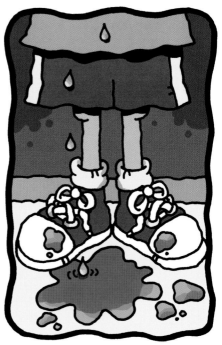

what cleaning supplies and rags were meant for. Give your children the inspiration to create science magic. The resulting laughter you will share is more important than a sparkling floor.

Teachers will enjoy this book as many of the activities can be safely done in a classroom setting. There are no expensive materials to purchase, and the experiments are bound to get an "Oh, cool!" response.

We have included a section on "things to do in restaurants when waiting for a meal" (WHERE'S DINNER? page 76). As survivors of restaurant outings with our own children, we can tell you that these diversions really work.

SAFETY FIRST!

It's great to have fun and experiment, but there are some simple rules you need to follow to make sure you won't hurt yourself or someone else. If you aren't sure if something is alright to do, ask an adult. Your adult supervisor— whether it's your parent, teacher, or babysitter—will be able to help you with the activities. Here are some guidelines you need to know before you get started.

Do's

1. Make sure an adult is with you while you do the experiments.
2. Tell an adult immediately if you or someone else is hurt in any way.
3. Have an adult read all the instructions for an experiment out loud with you before you begin any of the steps.
4. Have an adult gather all the materials you need for your activities.
5. Tie back long hair and roll up sleeves, so that they don't get in your way. Remove any loose jewelry, such as bracelets or necklaces.

6. Help clean up after you have finished each activity.
7. Wash your hands when you are finished with your activities.
8. Safely dispose of your experiments. Ask an adult for permission before dumping something down the sink or toilet.

Don'ts

1. Do not eat, drink, or taste any of the experiments or materials, unless the experiment says you can do this and an adult gives you permission.

2. Do not feed any experiments to other people or to animals either. Pets and children may get very sick if you feed them the wrong foods.

3. Do not gather your own materials or substitute other ingredients for the ones listed in the book.

4. If you have allergies to any food, do not do the activities that use the food.

5. Do not change the recipes in the book. Just because you add more of one ingredient does not mean the experiment will work better or faster.

6. Under no circumstances should you use the blender, food processor, hand mixer, sharp knives, or any other utensils or appliances unless an adult is helping you.

MATERIALS

Here is a basic list of things you will need. You probably have most of them in the kitchen already. See the project list for specific things you will need for that project:

- aluminum foil
- baking soda
- baking powder
- balloons
- bowls, large and small
- containers, large and small
- drinking glasses (plastic)
- food coloring
- funnel with narrow neck
- jars with lids (preferably clear plastic and large)
- magnifying glass
- masking tape
- measuring spoons, set
- measuring cup
- mixing spoons
- note pad
- paintbrushes
- paper plates
- paper towels
- paper
- pen
- pencil
- pie plates or shallow containers
- plastic bags
- plastic cups (small)
- plastic knife
- plastic wrap
- rags for cleaning
- ruler
- scissors (safety scissors)
- salt
- sponges
- strainer
- string
- toothpicks
- yogurt containers and lids, 6 oz. (170 g) or other small containers

LAB IN A BOX

Your parents can create a personal laboratory in a box for you. It's not expensive and will save you time in the long run as you won't have to look for various bits and pieces. Everything in this box should be child-friendly. Nothing sharp or dangerous should be put in the box. Don't put anything in the box that can attract flies or that needs to be in the refrigerator. An old large plastic container with a lid is a perfect "box" for storing science supplies. You might want to put the following items into your lab. As you get older and more responsible, your parents may add other things to this list.

✔ old shirt for lab coat (see page 5)
✔ plastic measuring cups
✔ measuring spoons
✔ string
✔ clear plastic jars with lids
✔ small plastic containers such as clean yogurt containers

✔ sponges
✔ rags for cleaning
✔ magnifying glass
✔ pencils
✔ food coloring
✔ small note pad
✔ masking tape
✔ small clear plastic bags

✔ safety scissors
✔ old spoon

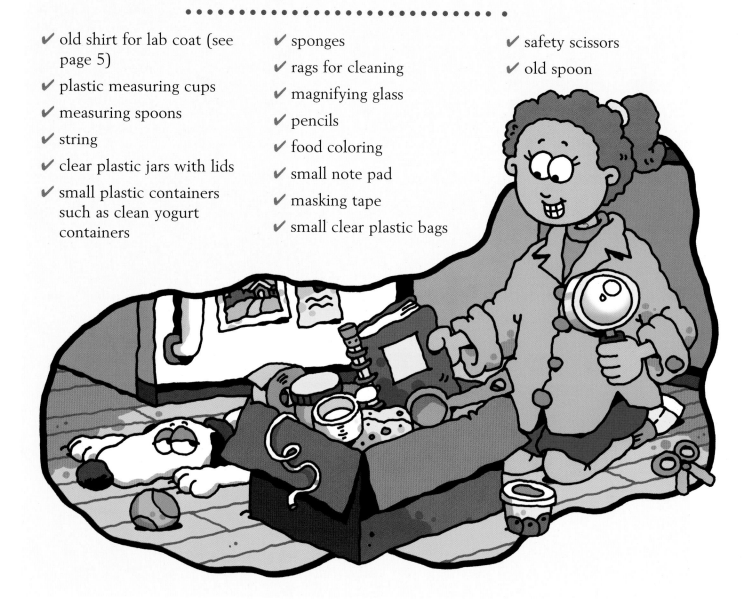

Fun Kitchen Science Fact: Medieval Toast

Do you like eating toast in the morning? If you wanted toast 1500 years ago, you couldn't just drop a slice into the toaster and wait until it popped up. Instead, you would have to stick your bread on a long fork and turn it over a large pit fire in the middle of the room.

Fun Kitchen Science Fact: Butter

Have you ever played the buttercup game? You hold a buttercup under someone's chin. If the yellow color of the flower is reflected and the person's chin looks yellow, you say, "You like butter." This saying comes from medieval times. People believed that butter made from the milk of cows that ate buttercups was yellower than other butter. In fact, butter made from the milk of cows that have been eating summer flowers — buttercups included — is yellower than other butter. Some of the flowers' yellow coloring passes into the milk and then into the butter made from the milk. The yellow color is prized by butter-eaters. Since the Middle Ages, people often have colored their butter with pigments (coloring) from plants and flowers to make it yellower.

GLOSSARY

When we introduce a new word, it will be in **bold** type. You will find its definition at the back of the book in the glossary.

Salted Popcorn

If you want to mix something up, you might put all the ingredients in a jar and give the jar a good shake. The up-and-down movement should combine all the foods and evenly mix everything together. That's what you'd think, but is that what really happens?

WHAT YOU NEED

- 1 cup (250 mL) table salt
- ¼ cup (60 mL) unpopped popcorn
- clear plastic jar with screw lid
- 4 or 5 large nuts such as brazil nuts
- ½ cup (125 mL) peanuts
- a few cashew nuts

WHAT YOU DO

1. Place the salt and popcorn in the plastic container and tightly screw on the lid.
2. Roll the jar on the table so the salt and popcorn are really mixed together.
3. Stand the container on the table or counter and gently bang the container up and down. Watch what happens to the popcorn.

4. Empty the container. Try the same steps again with a jar filled with large and small nuts. What happens to the large nuts when the container is banged on the table?

WHAT HAPPENED?

That's strange. The more you banged the jar, the more the popcorn rose to the top of the jar. Instead of mixing the two ingredients, you actually sorted them. In fact, the salt gets packed together under the popcorn. The tapping loosens the salt so that it moves down and around the popcorn. When the salt gets underneath the popcorn, it packs together tightly and the salt crystals form a solid layer that the popcorn can't get through. After you have given a few taps on the table, the popcorn has moved all the way to the top. The same thing happened with the large and small nuts: the large nuts came to the surface.

Pierced Potatoes

Sometimes when you try to put a straw in a juice box, the straw bends or breaks. This is pretty frustrating when you're thirsty. Can you imagine what it would be like to push a straw into something as hard and solid as a potato? Here's a neat trick that may come in handy the next time you are faced with a stubborn juice box.

WHAT YOU NEED
- plastic drinking straw
- raw potato

WHAT YOU DO

1. Hold a plastic drinking straw in one hand so that your thumb seals one end of the straw. Use your fingers to hold the middle of the straw firmly, so that it can't easily bend.
2. Quickly stab the open end of the straw into a potato. The straw should be at right angles (straight up and down) to the surface of the potato.
3. Remove the straw and try piercing the potato with the straw when you don't cover the straw end with your thumb.

WHAT HAPPENED?

You were able to pierce the potato when you sealed the straw with your thumb. When the top end of the straw was open, the straw couldn't pierce the skin of the potato as easily. Sealing the end of the straw traps the air inside. The air makes the straw stiffer, so that it can go into the potato without bending. The air takes up space and puts pressure on the potato. If the straw is open, the air just gets pushed out when you try to stick the straw into the potato.

Sun Tea

Iced tea was discovered by accident! In 1904 Richard Blechynden, a tea salesman from England, was at the World's Fair in St. Louis, Missouri. No one wanted to drink his hot tea because the weather was too warm. So he poured the tea over ice and created a new drink sensation!

WHAT YOU NEED

- large (quart or litre size) clear plastic jar or pitcher with a lid
- 2 tea bags
- measuring cup
- cold water
- wooden spoon
- plastic glass
- ice cubes (optional)
- lemon and sugar

WHAT YOU DO

1. Place the tea bags in the bottom of a clear jar or pitcher.
2. Add 2 cups (500 mL) of water to cover the tea bags and stir the mixture. Attach the lid firmly.

3. Place the jar or pitcher in a sunny spot and leave it until the same time the next day.

4. Pour your sun tea over some ice cubes in a glass and add lemon and sugar if you like them. Or put it in the refrigerator to cool. Enjoy your cool drink!

WHAT HAPPENED?

Inside the tea bags are dried tea leaves. When the tea leaves are placed in water, the substances in the leaves that can dissolve in water pass into the water to make a mixture called an **infusion**. When the sun heated up the water, more of the tea could dissolve in the water. Infusions can be made from any type of dried plant. You could try the experiment with herbal tea or with dried mint, basil, or sage.

DID YOU KNOW?

The first tea bags were made of silk. Tiny silk bags were filled with tea and dunked into hot water. Today tea bags are made from paper. The best tea doesn't make its way into bags at all, so if you want really good tea, you should buy leaves, have an adult help you make the tea, and strain it. Or you can make your own tea bags from pieces of cheesecloth tied closed with pieces of string.

Try This

Ice cubes for iced tea can be fun to make. Place a slice of lemon zest (the thin outer layer of the lemon skin) in each compartment of your ice cube tray. Then fill the tray with a mixture of 1 cup (250 mL) lemon juice and ½ cup (125 mL) sweet syrup and freeze.

A Penny for Your Thoughts

What are pennies made from? The shiny reddish metal you see when you look at a penny is **copper.** *Older pennies are copper all the way through. Nowadays, pennies are copper only on the outside and the insides are made from different, less expensive metals. Pennies start to look black after a while because the copper combines with* **oxygen** *in the air to make a layer of black* **copper (II) oxide.** *The black layer will come off when you do this activity!*

WHAT YOU NEED

- small jar or container
- ½ cup (125 mL) vinegar
- 20 copper pennies
- tablespoon (15 mL) table salt
- iron nail that will fit into your container
- steel wool or scouring pad
- strainer
- large bowl

WHAT YOU DO

1. Pour the vinegar into your jar or container. Add the salt and mix it well so that the salt dissolves in the vinegar.

2. Put the copper pennies into the vinegar. Let them sit in the vinegar mixture while you clean the nail.

3. Clean the nail with some steel wool or a scouring pad. Add the nail to the jar. Let it sit in the jar for about 5 minutes, until you see a change in the way the pennies and nail look.

4. Put a strainer on a large bowl and pour the pennies and vinegar into the bowl. Rinse off the pennies and nail under running water and look at them.

WHAT HAPPENED?

The vinegar-and-salt mixture cleaned the pennies and made them shiny. Vinegar contains **acid.** It is the acid that makes the pennies clean. The acid also reacts with some of the copper and makes a substance called **copper (II) acetate.** The iron from the nail reacts with the copper (II) acetate and the nail becomes coated with a little bit of copper.

DID YOU KNOW?

Salt-and-vinegar potato chips are not just regular chips dipped in vinegar and coated with salt. The vinegar would make them mushy. Instead, if you read the list of ingredients on the package label you will see a **chemical** called **sodium acetate.** This chemical is a powder that is used to coat the chips. When the sodium acetate dissolves in your mouth it makes vinegar!

Colorful Cabbage "Soup"

You probably don't think that your kitchen is a chemistry lab, but it turns out your kitchen and a lab have many things in common. Your kitchen and a lab both have sinks, counter space, and equipment for heating things. You also have a fair number of different chemicals in your kitchen. Some of those chemicals belong to two big groups called acids and **bases**. You may even have a few materials that can be used to tell which chemicals are acids and which are bases. Let's try making an acid-base tester.

WHAT YOU NEED

- adult helper
- small head of red cabbage
- plastic knife
- grater
- measuring cup
- 2 large bowls
- warm water from sink
- strainer or colander
- 2 or more clear plastic containers or glasses
- white vinegar
- baking soda
- foods and other things to test such as soda water, lemon juice, or yogurt
- jar with lid

WHAT YOU DO

1. Have an adult cut the cabbage in half.
2. Coarsely grate 1 cup (250 mL) of red cabbage.

3. Place the grated cabbage in a large bowl and pour in enough warm water to barely cover the cabbage in the bowl. Leave the cabbage mixture until the water is purple.

4. Place the strainer over the second bowl and pour the cabbage mixture into the strainer. Save the cabbage water.

5. Put ¼ cup (60 mL) of cabbage water into a clear container and add spoonfuls of vinegar until it changes color. What color does it turn?

Step 5. Adding vinegar turned the cabbage water pink. Plain cabbage water is on the left.

6. Put ¼ cup (60 mL) of cabbage water into another container and add some **baking soda**, a teaspoon at a time. What color does the cabbage water turn?

7. Try testing foods and other things in the same way to see if they make the cabbage water change color.

8. Save some of your cabbage water in a covered jar in the refrigerator to use in other experiments.

Step 6. With baking soda, the cabbage water on the left has turned blue.

WHAT HAPPENED?

The purple cabbage juice became blue-green when you added the baking soda to it. Baking soda is a base. Other substances that turn the cabbage juice blue-green also contain bases. When you added the vinegar to the cabbage juice, it became pink. Vinegar is an acid. Other acids or foods that contain acids will also turn the cabbage juice pink.

DID YOU KNOW?

Try testing beet juice, cranberry juice, and purple grape juice to see how they react. You may even get a reaction with certain colors of commercial food coloring.

ACIDS AND BASES

*You used acids and bases in this activity. You might have wondered what they are. Acids and bases are groups of chemicals. Acids have certain things in common with each other. Bases also have certain things in common. Acids, or foods that contain acids, such as lemon juice and vinegar, taste sour. Bases, such as baking soda or chalk, have a bitter taste. Many acids and bases are dangerous to taste, so we test them in other ways. To test whether something is an acid or a base, people sometimes use an **indicator,** a dye that changes to one color in an acid and changes to another color in a base. The purple cabbage juice in the Colorful Cabbage "Soup" experiment is an indicator.*

When you mix an acid and a base together in the correct amounts, a chemical reaction occurs which forms some new products that are not acids or bases. What you are frequently left with is a type of salt and water.

Now You See It, Now You Don't

Don't you wish you could make certain things disappear, like—say—your chores? This activity will not do the dishes, but it will show you how to make words or pictures vanish off the page. Then they will mysteriously reappear!

WHAT YOU NEED

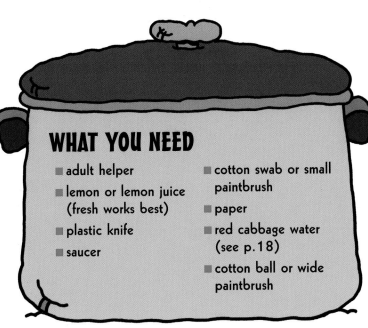

- adult helper
- lemon or lemon juice (fresh works best)
- plastic knife
- saucer
- cotton swab or small paintbrush
- paper
- red cabbage water (see p.18)
- cotton ball or wide paintbrush

WHAT YOU DO

1. Have an adult help you cut the lemon in half. Squeeze some juice from half a lemon onto a saucer. This is your "ink."

2. Dip the end of a cotton swab or a small paintbrush into the juice and use it to write a message or draw a picture onto a piece of paper.

3. Let the juice dry. The paper will appear to be blank.

4. Take a clean wide paintbrush or cotton ball and dip it in some red cabbage water. Paint over the paper that has the lemon writing.

WHAT HAPPENED?

Red cabbage juice is an indicator (see page 20). The acid in the lemon juice reacted with the cabbage juice and changed color, so you could see your message or drawing.

Fairy Rings

Mushrooms look like they would make good umbrellas for very tiny people. Did you know that they grow in rings called fairy rings? Mushrooms belong to a group of living things called **fungi**. **Yeast** is another type of fungus. Fungi have tiny seedlike particles called **spores**, which grow to make new fungi. Let's look at some spores.

WHAT YOU NEED

- fresh button mushrooms from store
- other fresh mushrooms from store, such as shiitake (optional)
- white paper
- sealable plastic bags
- magnifying glass

WHAT YOU DO

1. Remove the stems from a few mushrooms by bending them until the stems break free.

2. Place the mushroom caps, with the top of the caps facing up, onto a piece of white paper.

3. Put the mushroom caps and the paper into a plastic bag and seal the bag.

4. Leave the bag in a warm place overnight.

5. Open the bag, remove the mushroom caps, and look at the paper through the magnifying glass. You may see some brownish dust, which is really made of spores. If you don't see any spores, leave the mushrooms in the bag for a few more days until they ripen more, and then look again.

6. Try the same experiment with other types of mushroom, like shiitake mushrooms, if you have them.

WHAT HAPPENED?

When the mushroom caps were placed in the plastic bag and left in a warm place, the warmth and moisture in the bag caused the mushrooms to release their spores. The tiny brownish spores could just be seen with the magnifying glass. If you compare spores from a few kinds of mushroom, you will see that different kinds have different colors of spore. People who study fungi often use the color and shape of the spores to help them tell one mushroom from another. If the mushroom spores fell onto rich soil under the right conditions, they would grow into new mushrooms. Outdoors, mushrooms sometimes grow in a circle. This isn't because of fairies or anything magical—it's because mushrooms can get the most food from the earth that way.

Self-Inflating Balloons

Usually you blow up balloons with air from your lungs. Here's another way that may surprise you, using yeast. Dry yeast looks like tiny seeds, but you don't plant yeast in the ground. All the yeast needs is a little bit of warm water and some sugar and it will grow well, without any soil or sunlight. If someone in your household bakes bread or buns, you probably have yeast in your fridge.

WHAT YOU NEED

- balloons
- narrow funnel
- 1 tablespoon (15 mL) active dry yeast
- 1 teaspoon (5 mL) sugar
- measuring spoons
- measuring cup
- warm water from the sink
- helper
- ruler

2. Have your helper pour the yeast and the sugar into the balloon through the funnel. Then have the helper fill the measuring cup with warm water from the sink and carefully pour the water into the balloon until the balloon is full.

WHAT YOU DO

1. Place the bottom of a funnel into the opening of the balloon. You may need to stretch the opening of the balloon a little bit so that it fits.

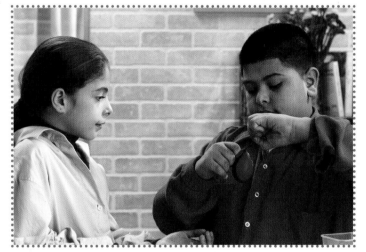

3. Remove the funnel from the opening of the balloon. Tie a knot in the balloon to keep the water-and-yeast mixture inside. Measure your balloon.

4. Place the balloon in a warm place and wait. Measure your balloon again.

WHAT HAPPENED?

The balloon started to get bigger. After a little while, the yeast began to grow. As it grows it expands or gets bigger and gets bubbly. Growing yeast gives off a colorless, odorless **gas** called **carbon dioxide.** This is the same gas that your body produces when you breathe. This gas fills up the balloon and the balloon swells up or inflates.

DID YOU KNOW?

Yeasts and mushrooms have something in common: they are both fungi. People use yeasts to make baked goods rise, but they also use them to make wine and beer. That's because another product of yeast, in addition to carbon dioxide, is alcohol. The earliest known yeast breads and beer were made by the ancient Egyptians about 4000 years ago.

Saltwater Daffy

All water is the same, right? It's wet. It's clear. It's great to drink when you are thirsty. When it freezes, it is wonderful to put into your tea or to skate on. If you like to go swimming, you have probably noticed that water is required for swimming too! Surprisingly, not all water is the same for swimmers. Let's look at some swimming eggs and see why.

WHAT YOU NEED

- 3 wide-mouthed large (about 16 ounces or 500 mL) clear plastic jars or glasses
- cold water
- 3 fresh raw eggs
- table salt
- measuring spoon
- mixing spoon
- turkey baster or spoon
- plastic cup

WHAT YOU DO

1. Almost fill two jars with cold water from the tap, but leave some extra room empty at the top. Fill a third jar halfway with cold water.

2. Use your hand to gently place a raw egg in the bottom of the first jar.

3. Place an egg in the second jar the same way. Add spoonfuls of salt to the second jar, gently mixing the water with a mixing spoon as you go, until the egg pops up to the top of the jar.

Step 3. The egg pops up in the second jar.

4. Place an egg in the third jar. Add salt to the third jar in the same way as in Step 3 until the egg floats to the top of the water.
5. Fill a plastic cup with fresh cold water and use a turkey baster to draw water from this cup. Hold the water-filled turkey baster against the side of the jar above the water line and slowly add the fresh water to the saltwater in the third jar. Try not to let the fresh water mix with the saltwater. The egg should stay suspended in the middle of the third jar. If you don't have a baster, gently pour the fresh water down the back of a spoon into the glass.

Step 5.a: Adding fresh water to the third jar.

Step 5.b: The egg hangs suspended in the water.

WHAT HAPPENED?

Each egg behaved differently depending on the kind of water in which it was placed. The egg in fresh water sank, but in water with enough salt in it, the egg floated. The salt changed the **density** of the water—it made the water thicker. When the water is less dense than the egg, the egg sinks. When the water is more dense than the egg, the egg floats. If you can adjust the density of the water to be the same as the density of the egg, the egg floats but doesn't come to the surface—it hangs suspended right in the middle of the jar. **Reminder:** Be sure to wash your hands and all equipment with warm soapy water after this experiment.

DID YOU KNOW?

Where does salt come from? Salt is produced all over the world. There are three main ways to get salt. Water from the ocean or salty lakes is dried out, leaving behind the solid salt. If the salt is under the ground, water can be pumped into the salt pockets. Then the salty water is pumped out and dried. In some places in the world, the salt deposits are so pure that the salt can be mined and dug out of the ground as solid salt.

Hard as a Rock

Have you ever pulled the brown sugar from the cupboard to put on your hot oatmeal, only to discover the yummy grains have turned into a solid brown block? It's not fun to chip away at the sugar for your breakfast. You can learn why this happens by doing a simple activity.

WHAT YOU NEED

- adult helper
- box of brown sugar
- plate
- measuring spoons
- 2 small clear sealable plastic bags
- apple
- plastic knife

WHAT YOU DO

1. Put some sugar from the box on a plate and look at it and feel it. Place 2 tablespoons (30 mL) of brown sugar into each plastic bag.

2. Leave the plastic bags open on a counter for several days.
3. Seal up one of the plastic bags with the dry sugar inside.
4. Have an adult cut a small slice from the apple. Place a small slice of apple in the second plastic bag with the dry brown sugar and seal the bag.

5. Look at and feel the outside of the bags each day to see if the brown sugar is hard or soft.

WHAT HAPPENED?

The brown sugar as you get it from the store is sealed in a plastic bag and has some moisture in it. It probably felt moist when you first touched it. When you left the bags of brown sugar open, the moisture in the sugar evaporated

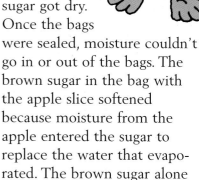

into the air. The water changed from **liquid** water to water vapor. The brown sugar got dry. Once the bags were sealed, moisture couldn't go in or out of the bags. The brown sugar in the bag with the apple slice softened because moisture from the apple entered the sugar to replace the water that evaporated. The brown sugar alone in the bag stayed hard.

DID YOU KNOW?

White sugar and brown sugar are the same, except that brown sugar crystals are coated in molasses. The molasses gives the brown sugar its brown color and adds to the flavor of the sugar. Sugar can be made from sugar cane or from sugar beets.

Rainbow Bursts

Contrary to popular belief, chocolate milk does not come from brown cows. Milk is white. But what would happen to milk if you added food coloring to it? Could you have found a new way to do art?

WHAT YOU NEED
- milk
- small bowl or container
- food coloring (a few colors)
- toothpick
- a little dishwashing liquid in a jar lid

WHAT YOU DO
1. Half-fill a small container with milk.

2. Add a drop of food coloring to the milk.

3. Dip your toothpick into dishwashing liquid. Just touch each drop of the food coloring on the surface of the milk with the tip of the toothpick and watch what happens.

4. Try the same experiment with several colors of food coloring at one time.

WHAT HAPPENED?

When the toothpick touched the surface of the milk, it released a little bit of the dishwashing liquid into the milk. This caused the food coloring to move away from the toothpick, which made a burst of color. Milk is mostly made of water, and water is made up of small particles called **molecules.** The molecules of water tend to stick together. When you dipped in the toothpick, the dishwashing liquid broke apart some of the water molecules and this caused the milk and colors to move and spread out.

DID YOU KNOW?

Milk is a complex mixture of different materials. It contains **fat, protein,** sugar, and water, as well as **vitamins** and **minerals** in small amounts. The sugar in milk is called lactose and it isn't as sweet as table sugar, so milk doesn't taste really sweet. Some of the vitamins in milk include riboflavin and vitamins B-12 and D. Many milk producers add Vitamin A to milk as well. Minerals in milk include **calcium,** magnesium, phosphorus, and zinc.

Bag It

Do you like ice cream? What's your favorite flavor? Many people like vanilla the best. Some children we know like really strange flavors—green tea, mango, licorice, or coffee ice cream. Whatever your choice, this activity is perfect for a hot summer day. Make sure an adult approves of any ingredient you are using before you add it to your ice cream.

WHAT YOU NEED

- adult helper
- measuring cup
- measuring spoons
- 1 pint-size (500 mL) resealable plastic freezer bag
- mixing spoon
- 1 gallon-size (3.7 L) resealable plastic freezer bag
- ice cubes, about 6 trays full*
- large bowl to hold ice
- 2 dishtowels
- rolling pin
- rock salt, 1 cup (250 mL)
- mittens or gloves
- freezer

*Note: You need the ice cubes to make about 6 cups (1.5 L) of crushed ice.

Basic Ice Cream Recipe

For each portion of ice cream, use the following ingredients:

½ cup (125 mL) of half-and-half**

½ cup (125 mL) heavy cream**

2 tablespoons (30 mL) sugar***

¼ teaspoon (1 mL) vanilla extract

EXTRA TREATS: 1 tablespoon (15 mL) of your favorite treat, such as crushed candies, small chocolate bar chunks, or chocolate chips

** Note: If you are allergic to milk, you can replace the milk and cream with soy milk or fruit juice. Use 1 cup soy milk or 1 cup of juice instead of the milk and cream.
*** Use artificial sweetening if you have diabetes.

WHAT YOU DO

1. Pour the **half-and-half, heavy cream,** sugar, vanilla, and the candy treat into the small freezer bag and mix with a spoon. Make sure you seal this bag tightly. Set the bag aside.

2. Empty the ice cube trays into the bowl. Get an adult to make crushed ice by wrapping the ice cubes in 2 dishtowels and smashing the ice with the rolling pin.

3. Empty the crushed ice into the gallon-size bag. Repeat the process—you probably have to make several more batches of crushed ice until you get about 6 cups (1.5 L) of crushed ice in the large bag.

4. Add about a cup (250 mL) of coarse rock salt to the ice in the large bag.

5. Place the bag with the cream mixture inside the large bag and seal the large bag.

6. Put on your warmest mittens or gloves and grab hold of the plastic bag. Shake the bag back and forth, up and down, and in every direction for about 10 minutes or more. Don't be too rough on the bag as you don't want to break any of the seals.

7. Place the bag in the sink and carefully open the outside freezer bag. Take the inside bag out of the large bag and rinse off the salty water from the inside bag's exterior. Put the ice cream bag in the freezer until the ice cream is frozen. This may take a few hours.

8. Open up the bag and you should have a frozen treat.

WHAT HAPPENED?

You made an icy dessert. Adding salt to the crushed ice makes the ice stay really cold. The sweet mixture in the inside freezer bag became cold enough to freeze. When you shook it all around, you kept the ice crystals that formed in the frozen treat from getting too big, so your ice cream has a nice, creamy texture.

TWO KINDS OF CHANGE

If the first thing that comes to mind when you think of "change" is the money you get for allowance, think again. In science, we have physical changes, which are changes in state or phase, and chemical changes.

How are these kinds of change different? Let's look at the changes that happen to water as an example. When water molecules (the very small particles that make up water) move closer together to form ice or farther apart to form steam, that is a change in state. The states are **liquid, solid,** *and* **gas.** *Changes of state can be reversed. Water can change from being liquid to solid when it freezes. Water can change back to a liquid again when it melts. It can even evaporate and become a gas in the air. If it gets cold again, the water that is a gas in the air can turn into a liquid again (rain), or a solid (snow).*

In a chemical change, something else happens. When water molecules break apart and their atoms combine to make oxygen and hydrogen molecules, which are different than the water molecules we had before, that is a chemical change. Chemical changes always make new materials that are different from what you started with, and you can't change them back to the way they were by heating or cooling them.

Milky Whey

"Little Miss Muffet sat on a tuffet, eating her curds and whey. Along came a spider and sat down beside her and frightened Miss Muffet away." If you have always wondered what a tuffet was, we can't help you, but this is how you make curds and whey.

WHAT YOU NEED

- ½ cup (125 mL) skim milk
- small plastic bowl or measuring cup
- measuring spoons
- white vinegar
- mixing spoon
- paper coffee filter
- strainer or colander
- large bowl to catch drips

WHAT YOU DO

1. Pour ½ cup (125 mL) of skim milk into a small bowl or measuring cup and add about 4 teaspoons (20 mL) of white vinegar. Stir the two ingredients together. You should see the milk start to change texture. Add more vinegar if it doesn't.

2. Leave the vinegar-and-milk mixture on the counter for about 10 minutes.

3. Put the coffee filter in the strainer or colander. Place the strainer over a large bowl (to catch the drips) and pour in your vinegar-and-milk mixture. Leave it until most of the liquid has passed through. The part that remains in the strainer is called the **curds.**

WHAT HAPPENED?

When the vinegar and milk were mixed together there was a chemical reaction. The vinegar caused proteins in the milk to clump together and form curds. Once the curds have formed, a liquid is left behind. It is called **whey.** Whey contains water, vitamins, minerals, and other proteins. About one-fifth of the protein found in milk is whey protein. The rest is **casein,** the protein which makes curds.

DID YOU KNOW?

Animals that make milk are called **mammals.** There are thousands of different mammals but humans drink the milk of only a few of them. Cow's milk is the most common type consumed by people, but people also drink the milk of sheep, goats, water buffalo, horses, camels, yaks, and reindeer. Of course most of us also started out drinking our mother's milk. Humans are mammals too!

Shake, Shake, Shake!

Have you ever wanted to experience time travel? If you could go back in time to the 1800s, you would find life very different than it is today. There were no supermarkets or cars or computers. The prepared foods that we take for granted today had to be made at home, even butter. Someone had to get up early in the morning to milk the cows. Then, after the cream had risen to the top of the milk, it was time to make the butter. Children in earlier years used a churn to make butter, but you can use any clean jar!

WHAT YOU NEED

- half-pint or 250 mL of whipping cream or heavy cream
- 2-cup (500 mL) plastic jar with a lid that seals to be watertight
- whisk or fork
- crackers
- butter knife

WHAT YOU DO

1. Pour the whipping cream into the jar and tighten the lid so that it seals.

2. Shake the jar back and forth for about 20 minutes.

3. Look in the jar from time to time. If the contents start to get too thick to shake, use a whisk or fork to stir them more.

4. When the contents start to form yellowish clumps, you are done. The solid yellow stuff you see is butter.

5. Spread some of the butter onto a cracker and taste it.

WHAT HAPPENED?

When you shake the cream for several minutes, it causes the little globs of fat in the cream to clump together with the protein and form solid butter. The liquid left behind is called buttermilk. When cows are milked, the fresh cow's milk has cream and milk all mixed together. The cream is less dense than the milk, so the cream rises to the top of the container, where it can be skimmed off. Skim milk is the milk left behind after the cream is removed. Homogenized milk is specially treated to keep the milk and cream mixed together so they don't separate.

DID YOU KNOW?

Margarine is not made from milk or cream. It is made from oils such as vegetable oil. The folks who make margarine add color and flavoring to make it taste more like butter. About 40 years ago, some places had laws to prevent margarine from looking too much like butter. Margarine was white, and customers had to add food coloring themselves to give it that "butter" color.

Brown Bagging

There are many different types of ingredient in the foods you eat, including sugars, proteins, and fats. Food scientists have developed tests to see whether a certain type of ingredient is in your food. Here is a simple test you can do to see if foods contain fat.

WHAT YOU NEED

- brown paper lunch bag or piece of brown kraft paper
- safety scissors
- pencil
- ruler
- 8 to 12 different food samples including butter, cooking oil, apple; you could use an avocado, nuts, cheese
- cotton balls or paper towel

WHAT YOU DO

1. Cut open the lunch bag with scissors, cutting off the bottom so that what is left is a large rectangle.
2. Draw lines to divide the paper into many equal-sized sections, enough so you have one for each food. Label each section with the name of the food you will test in that section.

3. Rub the butter on the square labeled "butter." Look at what happens to the paper in that section. Try holding the paper up to the light to see if the light shines through your butter test. Rub some oil on another square.

4. Rub the apple slice on the square labeled "apple." Look at what happens to the paper in that section.

5. Rub the other foods in their labeled squares. Use a cotton ball or piece of paper towel for the gooey ones if you want to. Then let the paper sit in a warm place to allow any wet sections to dry.

6. Compare the food tests from the other foods with the test squares of the butter and the apple.

WHAT HAPPENED?

Butter contains a lot of fat. The butter test square looks darker and glossier than the square where you rubbed the apple. Apples don't contain fat. When you first rubbed the apple slice on the paper, it may have turned the paper darker because the water in the apple made the paper wet. After a few minutes, that darkness faded as the paper dried. Foods that contain fat, such as oil and nuts, will look more like your butter test square. Foods that contain little or no fat will look like your apple test square. You may find that the fat samples make the paper translucent—they let light shine through the paper.

DID YOU KNOW?

When you look at raw meat or raw chicken, it's easy to see the fat. The thick white layers on meat and the yellowish soft material on chicken are fat. But what about nuts? All nuts contain oils, which are simply fat in a liquid form. Certain kinds of peanut butter have a liquid topping of oil, which is the oil that separated from the ground peanuts. Coconut meat contains coconut oils, which are added to things like cookies and crackers.

Salty Sidewalks

Why do people sprinkle salt on their outdoor stairs or sidewalks in the wintertime? Do they think that the concrete needs some flavoring? Actually, there's a simple reason for this odd behavior: salt makes snow and ice melt. Have you ever been ice-skating on a frozen pond or lake in the winter? Freshwater ponds will freeze if the weather gets cold enough, but saltwater usually won't unless it gets really, really cold. Why is that? What has the salt got to do with whether it freezes or not? Let's see.

WHAT YOU NEED

- adult helper
- two small (3 oz, 89 mL) plastic or foam drinking cups
- water
- measuring spoons
- salt
- spoon
- pencil
- freezer
- ice cubes
- 2 containers of cold water
- 2 pieces of string
- salt in shaker

WHAT YOU DO

1. Pour water into two small cups until they are about two-thirds full. Add a tablespoon (15 mL) of salt to one of

the cups and mix it well. Mark the cups so you know which has the salt.

2. Put both cups into the freezer. Check them every ten minutes to see which one freezes first.

3. Meanwhile, get an adult to help you take 2 ice cubes from the freezer. Put each ice cube in a container of cold water. Place a piece of string on top of each ice cube. Sprinkle one ice cube and string from above with a few shakes of salt.

4. After a few seconds, try to lift each ice cube with its string.

WHAT HAPPENED?

When you checked the cups in the freezer, you found that the water without the salt in it froze first. Fresh water freezes at a higher temperature than saltwater. So even though the saltwater in the freezer is the same temperature as fresh water, it didn't freeze as quickly, or maybe didn't freeze at all, depending on how cold your freezer is. The salt lowered the water's freezing point. To make the saltwater freeze, it has to get colder still. In the second experiment (Step 3), when you sprinkled salt on the ice cube, the salt made some of the ice melt and it also made the ice colder. The salt didn't get under the string, so the ice there stayed frozen. The string got damp from the water, and when it got cold enough, the string refroze and stuck to the ice cube so you could pick it up.

The fresh (unsalty) water froze first.

DID YOU KNOW?

Many people now use other chemicals or substances besides rock salt to keep their walkways clear in the wintertime. Sand and calcium chloride are safer for the plants in your yard than rock salt is, and they also melt ice. Before you help your parents to salt your walkways, ask them to check to see what else is available in your area.

Fizzies

Let's play food detective. You may have seen someone adding baking soda or **baking powder** to a cake, bread, or cookie recipe. But why did they do it? What do these two ingredients do, anyway?

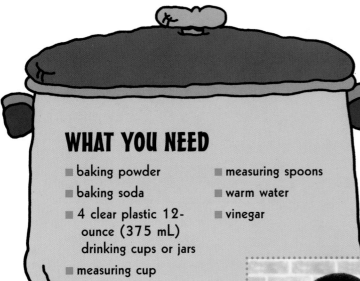

WHAT YOU NEED

- baking powder
- baking soda
- 4 clear plastic 12-ounce (375 mL) drinking cups or jars
- measuring cup
- measuring spoons
- warm water
- vinegar

WHAT YOU DO

1. Put ½ cup (125 mL) of warm water into each of the first two cups. Put ½ cup (125 mL) of vinegar into the third and fourth cups.

Step 1. Pouring vinegar.

2. Add 1 teaspoon (5 mL) of baking soda to the first cup of water and 1 teaspoon (5 mL) of baking powder to the second cup of water. Watch what happens.

Step 2. Adding baking powder to water.

3. Add 1 teaspoon (5 mL) of baking soda to the first vinegar cup and 1 teaspoon (5 mL) of baking powder to the second vinegar cup.

Step 3.a: Baking soda and vinegar.

WHAT HAPPENED?

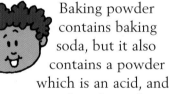

When the baking soda was added to the warm water it didn't fizz, but when it was added to the vinegar it fizzed like crazy. The baking powder fizzed in the water and in the vinegar. On the side of the box of baking soda, it may say it contains sodium bicarbonate or bicarbonate of soda, which are the chemical names for baking soda. On the side of the baking powder box, it says that it contains baking soda, as well as some other ingredients. Baking soda is a base (see ACIDS AND BASES, page 20), and it reacts to give bubbles of a gas called carbon dioxide when it is mixed with an acid like vinegar. Baking powder contains baking soda, but it also contains a powder which is an acid, and some cornstarch to keep the powder dry so that it doesn't form bubbles in the box.

Step 3.b: Baking powder and vinegar.

DID YOU KNOW?

Baking soda and baking powder are both **leaveners,** which means the gas bubbles they release cause breads, cakes, and cookies to rise so they end up being light and airy. When combined with an acidic ingredient like buttermilk, yogurt, or molasses, the baking soda releases the gas bubbles that cause the dough or batter to rise. Baking powder doesn't need to have something acidic in the dough or batter to work, because it already contains an acid as well as a base. Yeast also is a leavener (see SELF-INFLATING BALLOONS, page 25).

Playing with Dough

Many things in your kitchen can be fun to play with. A mixture of cornstarch and water makes an interesting concoction to push around on a tabletop. A little flour and water will work as a glue to make pieces of paper stick together. Cookie dough is fun to mold and shape, but our favorite mixture made in the kitchen is squishy play-dough.

WHAT YOU NEED

- 2 cups (500 mL) flour + some extra
- ⅔ cup (165 mL) salt
- measuring cup
- large mixing bowl
- 2 tablespoons (30 mL) powdered tempera paint (optional)
- measuring spoons
- wooden mixing spoon
- 2 tablespoons (30 mL) vegetable oil
- 1 cup (250 mL) water
- a few drops of vanilla extract (optional)
- cookie cutters (optional)
- can (optional)
- garlic press (optional)
- plastic bag to store dough

WHAT YOU DO

1. Pour 2 cups (500 mL) of flour and ⅔ cup (165 mL) salt into a large mixing bowl. Save some flour for later. Add the tempera paint now if you want colored play-dough. Use a wooden spoon to mix everything together.

2. Add the vegetable oil and some of the water (and vanilla if you want to) and mix everything with the spoon.

3. Keep adding water a little at a time until the dough is smooth but not sticky. If the dough gets too sticky, add more flour.

4. Dust your work surface with flour. Scoop the mixture out onto the work surface. Knead the dough by folding it over and pushing it flat for a few minutes until it is smooth and stays together in a ball.

5. Use your dough to make interesting sculptures. You can roll the dough out with the side of a can or press it flat with your hands. If you want to, use cookie cutters to cut out shapes or put the dough into a garlic press and make stringy "hair."
6. When you're finished playing, store your dough in a plastic bag in the refrigerator.

WHAT HAPPENED?

You made really neat dough! The flour and water helped it all stay together and the salt and oil kept it smooth and flexible. The dough will keep for quite a long time in the fridge as long it is sealed up in plastic. It is too salty to grow mold, because mold won't grow well on really salty food. The vanilla gives the dough a pleasant smell.

DID YOU KNOW?

When people make real dough instead of playdough, they want it to rise before it is baked, so that it will be light and fluffy. They use yeast (see SELF-INFLATING BALLOONS, page 25), which releases carbon dioxide gas as it grows and makes the dough rise (increase in size), or other leaveners.

Sally Sells Sea Salt by the Seashore

If you bought some sea salt from Sally, how would you know it was really sea salt and not sugar? Could you tell by just looking at it? How could you find out?

WHAT YOU NEED

- sea salt
- measuring spoons
- brown or black paper
- magnifying glass
- sugar
- table salt
- citric acid (sour salt)
- 5 shallow plastic containers or metal pie plates
- measuring cup
- warm water
- pencil
- paper for labels
- sea water (optional)

WHAT YOU DO

1. Place a teaspoon of the sea salt on a piece of dark paper. Take a close look at the salt, using a magnifying glass. What do the grains of salt look like?

Are they all the same, or are they different shapes and sizes?
2. Look at the sugar, table salt, and **citric acid** the same way.
3. Place several tablespoons of sea salt in a shallow container or pie plate and add ½ cup (125 mL) of warm water. Label the container so you know what is in it. Swirl this around to mix and leave it in a warm place for several days.
4. Repeat Step 3 with the table salt, citric acid, and sugar.
5. If you can get some sea water, place ½ cup (125 mL) in a pie plate, label it, and leave it to evaporate also.
6. After several days, all the water probably has disappeared from the containers. Use your magnifying glass to examine what remains.

WHAT HAPPENED?

As the water evaporated into the air, the dissolved substances started falling out of the water and forming crystals. A **crystal** is a solid that has a special, regularly repeating shape that grows out of the particles that compose the substance. Each substance made its own special kind of crystals. The sea salt and table salt crystals looked more like each other than they looked like the citric acid or sugar. If you try the experiment again, you will find that each substance forms crystals of the same shape again. If you studied them, you probably could learn to recognize each substance by its crystals.

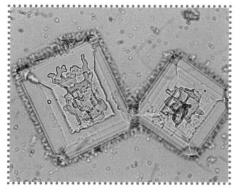

Photograph of a salt crystal, taken through a microscope.

Photograph of a sugar crystal, taken through a microscope.

Carrot Top

If you grow a plant from a seed, the roots grow down and the leaves and stem grow up towards the light. Here's a way of growing a plant that doesn't require a seed. It uses a food you may already have right in your refrigerator.

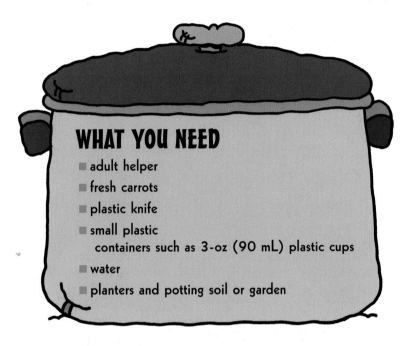

WHAT YOU NEED

- adult helper
- fresh carrots
- plastic knife
- small plastic containers such as 3-oz (90 mL) plastic cups
- water
- planters and potting soil or garden

WHAT YOU DO

1. Have an adult cut the lower part off a few carrots, about ½ inch (1 cm) down from the top. Trim off most of the leaves if there are any, but keep some of the green stems. (The experiment can be done with carrots that don't have leaves also.)

2. Place the tops of the carrots in the small plastic containers, with the cut side down. Add water to about ¼ inch (5 mm) depth.

3. Put the containers in a sunny place and watch what happens. Never let your carrots dry out; add water so the water level stays the same.

4. When the carrots have long roots, plant the carrots in the garden if the weather is warm enough, or indoors in a planter. Gently cover the roots with potting soil. The green leaves and a little bit of the carrot should stick out above the soil.

WHAT HAPPENED?

After a week or two, the carrot tops began to sprout new green leaves from the top and white roots from the bottom. This method of growing plants gives you new carrots from parts of the parent vegetable. Unlike the avocado pit in the TREE IN A JAR activity (page 54), the carrot top is not a seed. It is a root with some leaves.

DID YOU KNOW?

Wild carrots have been around for over 5000 years. Some of the earliest carrots weren't orange; they were purple. If Shakespeare were to have eaten a carrot, he would have had a yellow one, which was the color most varieties of carrot were in the 1500s. Scientists tried to make a better carrot, and created three basic kinds: red, yellow, and gold. The orange carrot we know and love today was developed by crossing the red with the yellow variety.

Which Way Is Up?

Turn a book upside down and see if you can read the words. Pretty confusing, right? Now bend over and look at the world upside down. That's even weirder. It's easy for us to figure out which way is up or down, but how about plants? Does a newly sprouted seedling know in which direction it is supposed to grow?

WHAT YOU NEED

- 4 or 5 beans or peas for sprouting*
- paper towels
- tall container or plastic glass
- water
- potting soil (optional)

*Note: You can get beans and peas for sprouting from a garden supply store. Scarlet runner beans work well. Or else try sprouting mung beans or chick peas, or even pinto beans.

WHAT YOU DO

1. Wrap the beans or peas in 2 pieces of paper towel and place the paper towels in the bottom of a small glass.
2. Dampen the paper towels with water. The towels should be completely soaked, but there shouldn't be extra water in the bottom of the glass. Pour out the extra water, if any.

6. Once the shoots and leaves have grown over the top of the container, place the container on its side.

7. Wait for two or three days to see what happens.

WHAT HAPPENED?

The beans or peas sprouted and the plants began to grow. Beans and peas grow quickly, so you probably didn't have to wait too long before you had a plant. Once the container was placed on its side, the plants began to grow at an angle to the container, as if they were drawn up to the ceiling. Plants have an ability to tell which way is up and which way is down. Even if you

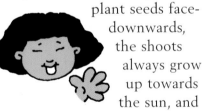

plant seeds face-downwards, the shoots always grow up towards the sun, and the roots grow downwards towards the earth. This response is called a **geotropism.**

3. Place the wrapped beans or peas in the upright container in a warm place and check them once a day. Add water, if needed, to keep the paper towels damp.

4. When the beans begin to sprout, arrange the paper towels so that the roots and beans stay moist and the shoots can grow up out of the container. Put them in a sunny spot and keep watering them.

5. If you like, you could plant the seeds in potting soil in the plastic container.

Tree in a Jar

In an old movie called Oh God, the character playing God was asked if he ever made any mistakes. He replied, "Yes, avocados. I made the seed too big." As it turns out, this "mistake" is something you can use.

WHAT YOU NEED

- adult helper
- ripe avocado that has not been refrigerated
- plastic knife
- paper towel
- 3 or more toothpicks
- plastic jar with an opening slightly wider than the avocado pit
- water

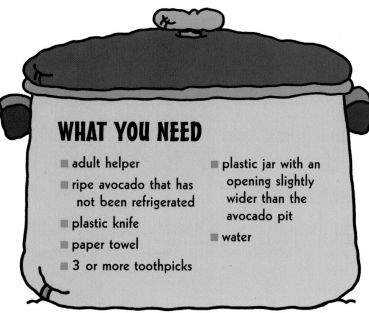

WHAT YOU DO

1. Have an adult help you cut open the avocado and remove the pit. Wash the pit and use a paper towel to wipe off any avocado goo from the pit, so your seed is clean and shiny. You don't have to take off the brown pit covering.

2. Get an adult to help you evenly space three toothpicks around the center of the avocado pit and push them about ½ inch (1 cm) into the pit.

These will hold the avocado in place over the water.

3. Gently balance the fat end of the avocado pit over the opening in the jar. The pointed part of the pit should face up. The toothpicks should stick out far enough to keep the avocado from falling into the jar. Add more toothpicks if you need to.

4. Pour water into the jar up to the very top of the jar, so the water touches the bottom of the avocado.

5. Set the jar in a warm place out of the sun and look at the bottom of the pit every day over the next few weeks. Add water to keep the water level touching the bottom of the avocado at all times. In two to six weeks, your pit will begin to grow roots and then a stem.

6. When the stem of the plant is about 6 inches (15 cm) tall, cut it back to about 3 inches (7.5 cm) to make the tree bushier.

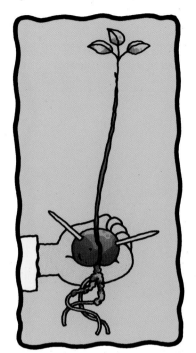

7. Once the stem of the tree grows back to about 6 inches (15 cm) tall, you can plant it in earth. Be sure the roots are covered, but leave the top of the pit sticking out of the earth. If you live in a warm place like California, you can plant your avocado tree outside. If you live in a cooler place like Seattle, you can grow the tree inside in a planter.

WHAT HAPPENED?

An avocado pit is a seed. After a few weeks, roots began to sprout from the bottom of the avocado pit. Soon, the pit split in half and a shoot began to grow from inside the pit. In the right climate, your avocado might grow into a tall tree. In a planter, it will grow to be a pretty indoor plant. Avocados are very good for you. Avocados are not considered vegetables, but are instead a type of fruit. Avocados contain a lot of oil. Try testing an avocado using the technique in BROWN BAGGING, page 40.

DID YOU KNOW?

Avocados grow on enormous trees in hot places like Hawaii and Florida. Avocado trees can grow to be 60 feet (18 metres) tall. There can be 200 to 300 avocados on a single full-grown tree.

Popcorn on the Cob

Sometimes in the back of your cupboard you may have come across a potato or two sprouting long shoots. The potato may not have even been there that long, but it is growing anyway. Why is it that you can keep popcorn in a jar for years and it won't grow? Bet you'd much rather have a popcorn plant than a potato plant.

WHAT YOU NEED

- adult helper
- unpopped popcorn*
- paper towels
- water
- plastic container with lid

- potting soil
- small garden pots
- plant food

*Note: Use the kind from a jar, not the popcorn in a microwave bag or the one in a metal container.

WHAT YOU DO

1. Take about 12 unpopped popcorn kernels and place them on a single sheet of paper towel. Put the towel with seeds in a plastic container.

2. Wet the towel and fold the towel in half, covering the popcorn.

3. Close the container and leave it in a warm place for several days. Be sure to keep the towel damp.

4. The kernels should start to grow. When the kernels have opened and you can see white roots coming from the bottom of the seed, carefully pick up each sprouted seed and gently plant it in a small pot of potting soil, one seedling per pot, about ½ inch (1 cm) down from the top of the soil.

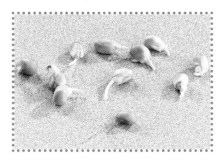

5. Keep the pots well watered and feed the seedlings the plant food (get an adult to help you mix it if necessary) until the weather is warm enough to transplant them outdoors. If you don't have a garden, you can transfer the popcorn plant to a large container after it has grown 5 or 6 inches (12 or 25 cm) tall. Keep the plant well watered.

6. If conditions are right, your popcorn plant will grow into a tall stalk of corn and will one day grow several ears. These ears will not get as large as the corn you eat for dinner. They will only be about 6 inches (15 cm) long. When the stalk dries out, pull off the ear from the stalk and peel back the husk to reveal the corn.

WHAT HAPPENED?

The popcorn kernel is a seed. It has a hard shell and it does not sprout without water. The water seeps into the shell, causing it to swell up and split. These seeds begin a process called **germination:** The tiny beginnings of the plant inside the seed start to grow, and roots and shoots come from either side of the seed. The dried seeds are **dormant,** which is like being asleep. When you add the water, the seeds "wake up" and start growing into plants.

DID YOU KNOW?

There are different kinds of popcorn. Next time you are at the movie theatre, check out the popcorn in your bag before you eat it. You will see that some kernels pop in big fluffy shapes. These are "butterfly" varieties; they take up lots of space in a popcorn bag. There are also kernels with round heads. This type of corn holds butter or topping without breaking.

Racing Colors

Have you ever stuck a candy on your tongue and sucked off the color? What color was your tongue after you did this? Did you know that the bright color usually comes from chemical dyes? Here is a scientific way to learn a little about food coloring.

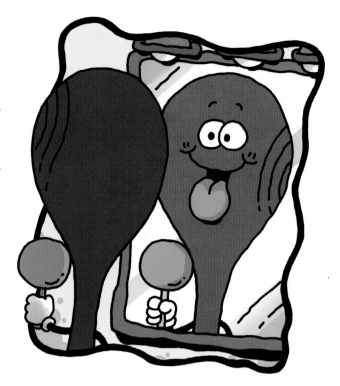

WHAT YOU NEED

- white coffee filters
- 2 or 3 colored candies or gumballs*
- 2 clear plastic drinking glasses or narrow jars
- measuring spoons
- water
- food coloring, a few colors

*Note to parents: Children age 5 or younger should not chew on gumballs as they could choke.

WHAT YOU DO

1. Fold a coffee filter to make a cone. If it is already a cone-shaped filter, you don't have to do anything.
2. Place a colored candy or two or a few gumballs inside the tip of the cone.

Step 1. Folding the filter into a cone.

3. Put a little bit of water in the bottom of the plastic drinking glass—about a teaspoon (5 mL). Place the coffee filter cone in the glass so that the tip of the filter just touches the water.

4. Wait and watch what happens.

5. Mix a few drops of several colors of food coloring together in a second glass and add about ½ teaspoon (2.5 mL) of water.

6. Fold another coffee filter into a cone and put it in a container so the tip just touches the food coloring. Wait and watch what happens.

Step 4. The gumballs are inside the filter, and the food coloring is rising in the filter.

Step 6. Food coloring is rising in the filter and starting to separate into rings.

WHAT HAPPENED?

In both parts of the experiment, the water begins to move up the coffee filter as the filter absorbs (takes in) some of the water. As the water moves up the filter, it pulls along the colored dye that coats the candies or the dye that is in the food coloring. Different colors of dye move at different speeds, so some dyes move ahead faster and they reach higher up on the filter. The dyes get separated in the filter.

DID YOU KNOW?

If you tried the same experiment again with colored candies or gumballs of the same color from different candy manufacturers, you might see different colors on the filters. The colors might travel at different speeds along the filter paper and end up in different positions than before. This happens because the dyes used to make candies red, for example, can vary from one manufacturer to another. Food scientists test dyes using a process called paper chromatography, which separates the dyes on long strips of paper.

Taking Stalk

Do you like using a straw to slurp up your favorite drink? How do you think plants get their liquids? Can they use straws, too? You may never have seen a plant take a drink, but here's an activity to show you just what they do.

WHAT YOU NEED

- adult helper
- thick piece of celery
- plastic knife
- 2 small plastic drinking cups of the same size
- water
- food coloring
- white carnation* (optional)
- plastic container

*Note: Carnations with thick stems work best.

WHAT YOU DO

1. Have an adult use the knife to split the celery in half lengthwise partway, but do not completely split the celery into two pieces. Leave an inch or two (2 to 5 cm) at the top of the celery where it is not split.

2. Fill the two small cups about three-quarters of the way with water. The water must be high enough to cover the split ends of the celery. Place the glasses next to each other. Place one side of the cut celery stalk in one cup and the other side in the other cup.

3. Add blue food coloring to one cup and red food coloring to the second one.

4. After a few hours or overnight, check back on the celery.

5. Take the celery out of the water and have an adult cut across the celery in a few places. What do you see?

Step 5. Cut celery shows red and blue food coloring.

6. If you wish, cut a carnation so its stem is about 4 or 5 inches (10 or 12 cm) tall. Place the stem in a container with red or green food coloring. Leave it for a day or more. What happens to the carnation when the stem is left in the colored water?

WHAT HAPPENED?

The celery stalk is the stem of the celery plant. When you cut the celery across after it spent some time in the food coloring, you could see tubes filled with red or blue liquid. The liquid was drawn up through the stalk of the celery through strawlike cells in the stalk. The celery uses the strawlike **cells** to bring **nutrients** and water from the soil up the stem to the leaves after the roots bring the nutrients into the bottom of the plant. Even though the celery has been cut from its roots, the cells in the celery stalk still work to pull up water. The stem of the carnation also has tube cells that bring water up the stem. The carnation flower was colored by food coloring that traveled up the stem to the flower, along with the water.

DID YOU KNOW?

Roses come in many different colors. Most of the roses that people give for Valentine's or Mother's Day are grown in greenhouses. The growers add coloring to the water to make some of the roses vivid colors that don't occur in nature. The color gets pulled up the rose stems and into the flowers.

Rusting Apples

Apples make great snacks in a lunch bag, but have you ever unwrapped a cut-up apple at lunchtime, only to find that your shiny red apple, which started out with a glossy white interior, had turned into a brownish yucky mess? What happened to your healthy treat?

WHAT YOU NEED

- adult helper
- apple
- plastic knife
- 6 paper plates
- lemon juice
- vinegar
- milk
- water
- plain soda water (seltzer)
- 6 spoons
- pencil

WHAT YOU DO

1. Have an adult cut the apple into six slices. Place one slice on each of the six plates.

2. Sprinkle one slice of apple with lemon juice. Use a pencil to write the word "lemon" on the plate.

3. Sprinkle the second slice with vinegar, and write "vinegar" on this plate.

4. Sprinkle milk on the third slice. Put plain water on the fourth slice and soda water on the fifth slice. Do not put anything on the slice on the last plate. Label each plate.

5. Leave the plates on the counter for an hour. Then check your experiment.

WHAT HAPPENED?

The apple slice with the lemon juice and the ones with the soda and milk were still white. The slice that had nothing on it started to turn brown. The apple slice in the vinegar was the brownest of all. Apples turn brown when they are exposed to air because the oxygen in the air combines with some of the naturally occurring chemicals in the apple. Milk, vinegar, and lemon juice all contain different types of acid. Lemon juice contains citric acid and ascorbic acid (Vitamin C). Vitamin C is an **anti-oxidant;** it stops the reaction of the oxygen with the apple, so the apple with lemon juice didn't turn brown. Not all the acids work in the same way—in fact, some do the opposite and make the apple brown sooner than air alone does.

DID YOU KNOW?

Here's a fun activity you can do while waiting for your dinner to arrive at the table. Place a huge red apple on a piece of white paper and stare at the apple for about a minute. Take away the apple and look at the white paper. A green apple should seem to appear before your eyes.

Genie in a Bottle

When Aladdin rubbed his magic lamp, a huge genie appeared and helped Aladdin. But how did that genie get into the lamp in the first place? Perhaps this next activity will give you an idea of how truly difficult it is to get a genie into a lamp. Please note: a crumpled-up ball of paper inside a bottle will not grant you wishes. Sorry.

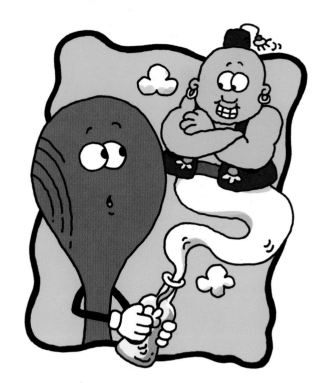

WHAT YOU NEED

- scrap of paper
- empty soda bottle
- helper

WHAT YOU DO

1. Crumple a small piece of paper into a small ball, slightly smaller than the opening of the bottle.

2. Place the paper ball on the countertop, keeping your chin just above the counter, and blow on it gently from one side. It should move easily along the countertop.

Step 2. Blow the paper from the side; it moves easily.

3. Have a helper hold the bottle so that the opening of the bottle rests on the countertop. Place the paper ball on the counter at the opening of the bottle.

4. Blow gently against the paper ball. Can you make it go into the bottle?

WHAT HAPPENED?

The paper ball moves easily when you blow on it, but it won't go into the bottle. That is because something is already in the bottle, even though it looks empty. What is in the bottle? Air! The only way that the crumpled paper ball will go into the bottle is if some of the air inside the bottle is allowed to come out. When you blow on the paper, you keep the air inside the bottle from leaving, so the paper won't go in.

Step 4.a: Trying to blow the paper ball into the opening of the bottle.

Step 4.b: The paper ball doesn't go into the bottle opening.

How Now, Green Cow?

What do cows and artists have in common? As it turns out, milk is something people can use to create paints. Here is another way of enjoying milk.

WHAT YOU NEED

- nonfat powdered milk
- measuring spoons
- measuring cup
- powdered tempera paints in a few colors
- warm water
- small plastic containers or jars with lids (one for each color + one more)
- mixing spoons
- paintbrush
- paper to paint on

WHAT YOU DO

1. Place 3 tablespoons (45 mL) of nonfat powdered milk in a small plastic container. Add one tablespoon (15 mL) of powdered tempera paint and stir the paint and dried milk with a spoon until it is all mixed together.

2. Slowly add water until the paint is a good thickness for painting, mixing with a spoon until the paint is smooth and even-colored.

3. Repeat Step 1 and Step 2 for each color of paint you want to make. Try blending the colors by mixing two different colors of tempera if you like.

4. In one container, place one tablespoon (15 mL) of powdered tempera and enough water to thin it to paint, but don't add any milk. Mix the water and tempera together with a spoon.

5. Try out all the colors you mixed. Use your paintbrush and paper to make a picture. Save some paint for printmaking (see DESIGNER FOODS, page 68).

WHAT HAPPENED?

You made milk-based paint. The tempera and dried milk combined to give a thicker paint than the tempera paint gives by itself. The proteins in the milk hold the paint together and make it very sticky. You can see this if you try gluing two pieces of paper together with some of the paint.

DID YOU KNOW?

Milk paint is one of the oldest types of paint around, but it is also one of the newest. Because many people nowadays have allergies to the chemicals in paints, they are starting to use milk paint. Milk paint was used in the past by families or traveling painters who made their own paints, but now it can be ordered from companies who specialize in this old-fashioned paint.

Designer Foods

Take a close look at some of the fruits and vegetables in your kitchen. They are good, healthy treats to eat, but what else can you do with them? Here's a cool way to use fruits and vegetables as stamps to decorate things. At the same time, you'll get to take a new look at some of your favorite treats.

WHAT YOU NEED*

- adult helper
- fruits and vegetables such as:*
  - apple · celery · broccoli
  - carrot · large onion
  - starfruit
- plastic knife
- magnifying glass
- clean kitchen sponge
- safety scissors
- milk paint from page 66 or other paint, such as tempera paint
- jar lids or shallow containers
- spoons
- newspaper
- papers you want to decorate

*Note: If you can't get these fruits and vegetables, experiment with other ones that you have.

WHAT YOU DO

1. Have an adult help you cut the fruit in half, and cut sections from the carrot and from the stems of the celery and broccoli. Pull off some celery leaves. Cut off a few broccoli flowerets too.

2. Use a magnifying glass to take a close look at the cut fruits and vegetables. What can you find inside a carrot or broccoli stem?

3. Have an adult cut a large onion in half. Use the half of the onion that has the roots attached as your stamp. This will make rings.

4. If you can find starfruit in your grocery store, you can use it to make perfect "stars" for printing. Ask an adult to slice this fruit across in various places, and you will have stars of different sizes for printing.

5. Have an adult cut the sponge into several pieces, one for each color of paint. Put a small piece of sponge in a shallow container or jar lid. Put some paint on the sponge with a spoon.

6. Press various parts of the fruits and vegetables into the paint on the sponge and stamp them onto paper, paper bags, or even notebooks. Try various colors and designs. Test them out on newspaper first to see if you like them. Remember not to decorate any walls or kitchen surfaces!

7. Can you tell from the images you made what kind of plants they came from? See if a friend can guess.

WHAT HAPPENED?

When you dipped them in paint and pressed them onto paper, you could see the patterns that each fruit or vegetable stamp created. Inside the apple, you found seeds held in a central core. Root vegetables like carrots and potatoes are solid inside. Leaves like the tops of celery make a feathery pattern. A cut across a stem may show the tube cells (see TAKING STALK, page 60). Inside

the onion you can see the layers that protect the central plant. Each of the plant parts makes a stamp with a different pattern.

DID YOU KNOW?

Stamping has been around for thousands of years. In ancient times, people wore special rings that had carved designs. The surface of the ring was painted, and the ring was stamped on a surface like clay or early forms of paper. These were called "signet stones." They were used instead of a signature to identify a person. It was easy to fake a signature, but really difficult to fake a stamp.

Name That Food

Hold an orange in your hands. Take a deep breath. It smells like an orange. What do you think is inside the orange? Juice, maybe some pits, some white pith, and that's pretty much it. But what do you think is inside some packaged foods? Here's a game to play with your family and friends to find out.

INGREDIENTS
rice vinegar
honey
canola oil
xanthan gum
artificial color
carrageenan

WHAT YOU NEED

■ labels from packaged foods such as:
- • canned tomatoes • macaroni and cheese
- • pudding mix • crackers • fruit drink mixes
- • dried potato mixes • ice cream • ketchup
- • bread • tomato sauce • salad dressing

■ safety scissors

■ pencils and paper

■ friends

WHAT YOU DO

1. Have an adult help you cut off or remove the labels of each of the foods.

2. Cut the ingredients list out of the label. Don't leave any part of the name or other information that will identify what it came from. Write a number on the back of each label. On a separate piece of paper write the food's name

and the number you assigned to it. This way you will be able to match the food with the number later on.

3. Get some friends together and pass around the ingredients lists. Have each person try to guess what product he or she thinks the list is from.

4. Check the answers. How many people guessed correctly and could identify the products?

WHAT HAPPENED?

Sometimes it is hard to tell what a food is because there are many things added to foods. Some things are added to keep the foods from spoiling. Other things are added to make the foods taste better. Sucrose and fructose are kinds of sugar. Some chemicals, like potassium phosphate and sodium phosphate, are used to keep foods from spoiling. Riboflavin, thiamine, and biotin are vitamins. You can look at books in the library or do a search on the Internet under *food additives* to learn about why different additives are used in foods.

DID YOU KNOW?

You know what strawberries and bananas taste like, but did you know that you can taste that same flavor by eating a chemical? Scientists can recreate many of the distinctive flavors and odors of your favorite

foods in the laboratory. Although these chemicals may not taste precisely the same they can sometimes fool you.

Artificial (manmade) **flavors** can be easier and less expensive to use for foods made in large quantities.

Birds of a Feather

Unless you live downtown in a big city, you can probably look out the window in the morning and see birds flying by. There are all kinds of different things in your kitchen that birds will eat. Here's one way to make a handy bird feeder. If you start feeding birds in the fall you should keep your feeder stocked so the birds don't get hungry when winter comes!

WHAT YOU NEED

- string
- pinecones
- spoon
- vegetable shortening or lard
- birdseed, about 2 tablespoons (30 mL) for each pinecone
- shallow bowl
- pencil and paper

WHAT YOU DO

1. Thread a piece of string around the stem of a pinecone and tie it in a knot. You will use the string to hang up your feeder.

2. Use a spoon to smear a pinecone with vegetable shortening or lard. If possible, the shortening or lard should go inside the bumpy parts of the pinecone (the scales).

3. Pour some birdseed into a bowl. Roll the pinecone in the birdseed, pressing down with your fingers to make the seed stick to the shortening or lard on the outside of the pinecone. Put birdseed onto the pinecone until all the shortening is covered.

4. Tie the string ends to form a loop so you can hang up your bird feeder. Have an adult help you place your feeder somewhere out of the reach of cats, otherwise this will become a cat feeder. You could hang it from a tree.

5. Visit your feeder often to see if any birds come by. Get a bird book from the library and try to identify your visitors. Keep a record of who visits your bird feeder and when they visit. Do your visitors change at different times of the year?

WHAT HAPPENED?

You made a bird feeder. If you hang it in a safe place, the birds will come after a while. You can see what types of bird are eating at your feeder, and maybe you will spot an unusual or rare bird!

DID YOU KNOW?

Fewer than 10,000 species of bird have been identified in the world. Every year a few species become extinct or disappear, and often new species are found and identified. People who like to go bird-watching often keep a record of which species they have seen and identified. Even if you write down every different kind of bird you see, you will have to work pretty hard to sight as many birds as Phoebe Snetsinger, a Missouri woman who sighted more than 8000 different species of bird in her lifetime. She traveled all over the world to look for birds.

The Three Bears

Maybe the three bears were onto something with their porridge. One bear thought his porridge was too hot. The second found hers too cold. But the baby bear thought his porridge was just right. What scientific principle did the baby bear know that his parents didn't know?

WHAT YOU NEED*

- adult helper
- ice cream
- small bowl
- small chocolate bar
- freezer and refrigerator
- slice of pizza
- aluminum foil
- plastic knife
- spoon

*Note: Parents should approve the items to be eaten. Don't do the experiment with any foods to which you are allergic.

WHAT YOU DO

1. Place a small scoop of your favorite ice cream in a bowl and leave the bowl on the counter for about a half-hour.

2. Take a small chocolate (one square from your favorite kind of treat) and place it in the freezer for about an hour.

3. Have an adult cut a slice of room-temperature pizza in half. Wrap half in foil and place the half in the fridge for about an hour.

4. Now for the fun part—the taste test. Take a bite of the room-temperature pizza, then a bite of the cold pizza from the fridge. Which tastes better?

5. Remove the chocolate square from the freezer and take a bite. Now bite into the room-temperature chocolate. Which tastes better?

6. Eat several spoonfuls of ice cream from the freezer. Then taste the room-temperature ice cream. Which tastes better?

WHAT HAPPENED?

The temperature of the food changed the way the food tasted. Frozen ice cream tastes better than melted ice cream. Frozen chocolate has no taste at all compared to the chocolate at room temperature. Cold pizza tastes like cardboard compared to warm, gooey pizza. Usually the warmer food is, the more intense the flavors, because when food is warm, more of the substances that give food flavor pass into the air, and you smell them, which enhances the flavor of the food. Certain types of food, like ice cream, are made to be enjoyed at cooler temperatures and just taste weird if you heat them.

DID YOU KNOW?

Prepared foods must be kept at certain temperatures to avoid food poisoning. Hot foods must be kept at 140°F (60°C), and foods that are frozen must be kept below 40°F (4°C). People have made these rules because there are germs or bacteria that can grow rapidly if food is between 40°F and 140°F; they can cause upset stomachs and food poisoning.

Where's Dinner?

Don't you hate it when you are really hungry and dinner's not ready? Your mom won't let you have a snack because it will spoil your appetite, and you are sitting patiently at the table in the hope that a piping hot meal will soon be there. Or even worse, you have to behave yourself in a restaurant until you are served. What's a kid to do? Here are some activities you can do that will keep you and your growly tummy occupied until the food arrives.

1. THE SPOON TRICK

This silly trick will help keep your mind off food.

WHAT YOU NEED

■ a spoon

WHAT YOU DO

Lick the inside of the spoon and see if you can get it to stick to the tip of your nose. Get everyone else at the table to try the same thing.

2. SALT-AND-PEPPER MAGIC

Think you can separate the pepper from salt, without touching either spice? Here's a nifty bit of science magic.

WHAT YOU NEED
- salt
- coarse pepper
- plastic disposable pen
- plate
- scarf or some material

WHAT YOU DO

Sprinkle some salt and pepper onto a small plate. Rub the pen between some pieces of material, like a scarf, wool mittens, even the t-shirt you are wearing. This will create a static charge. Hold the pen just slightly above the salt-and-pepper mixture and slowly rotate the pen over the plate. Watch the pepper jump onto the pen. Rubbing the pen gives it an electric charge, which attracts the pepper.

3. RAISIN ELEVATORS

How about getting some dried fruit to help entertain you? Watch these things boogie.

WHAT YOU NEED
- adult helper
- glass of soda water or clear soda pop
- about 5 raisins
- butter knife
- spoon

WHAT YOU DO

Pour some soda water into a glass. Have an adult cut the raisins in half and drop the halves into the fizzy liquid. Wait a few seconds and watch what happens to the raisins. Soon bubbles start picking up the treat and giving them rides to the top of the glass. As the bubbles pop, the raisins go back to the bottom of the glass. Take a guess as to which raisin will catch the next ride. Use a spoon to scoop out the raisins before drinking the soda.

Glossary

acid: a sour substance that forms water and a salt when it is mixed with a base

anti-oxidant: substance that keeps other substances from combining with oxygen

artificial flavors: manmade substances added to food to imitate certain flavors

atom: the smallest particle of an element

baking powder: a white powder that contains baking soda and an acid ingredient. Used to make dough rise in baking

baking soda: a white powder, also called sodium bicarbonate or bicarbonate of soda, used to make dough rise in baking

base: a bitter-tasting substance that forms water and a salt when mixed with an acid

calcium: a metal element which is part of bones, teeth, and sea shells

carbon dioxide: an odorless, colorless gas that is produced by plants, animals, and yeast

casein: a milk protein

cells: small units that make up living things

chemical: a substance made by a chemical process

citric acid (sour salt): a white powder made from the juice of citrus and other fruits, used as a flavoring for foods and drinks

compound: a pure substance that consists of two or more elements combined in a fixed formula. For example, water is a compound of hydrogen and oxygen

copper: a reddish-brown metal element

copper (II) oxide: a blackish-brown compound containing copper and oxygen

curds: the solid material formed when cheese is made from milk

crystal: a solid that has a special, regularly repeating shape that grows out of the particles that compose the substance

density: the amount of matter of a substance that is contained in a certain space

dormant: alive but in a resting condition

element: one of about 114 pure substances that consist of only one kind of particle. Single elements or numbers of elements combine to form other materials

fats: oily or greasy materials found in milk, meat, nuts, or in the seeds of plants. Fats make unglazed paper translucent

food additives: substances added to foods to make them more appealing, to keep them fresh, to help their nutritional quality, or for other reasons

fungi: group of living things that includes mushrooms, molds, yeasts.

gas: the fluid form of a substance (for example, air) that expands to fill a container and doesn't have a definite volume

geotropism: the movement or growth of a plant in response to gravity

germination: the process of sprouting from a seed, bud, or spore

half-and-half: a mixture of half milk and half cream

heavy cream (heavy whipping cream): whipping cream with a high butterfat content (between 36% and 40%)

ion: an atom or molecule that has an electric charge. Salts form ions when they melt or dissolve in water.

indicator: a dye substance that changes color in the presence of an acid or a base

infusion: a liquid extract made from soaking something in water

iron: a white metal essential for plant and animal life; iron is the most common metal on Earth

leavener: a substance such as yeast, baking soda, or baking powder. Leaveners are added to bread, cakes, or cookies to make them rise during or before baking

liquid: a fluid, like water, that has no definite shape but has a definite volume

mammals: warm-blooded animals that produce milk to nourish their young

minerals: elements or compounds containing metals found in the Earth, needed by plants and animals for proper growth

molecules: the smallest part of an element or compound that can exist freely and have the properties of that substance. Molecules are made of even smaller units called atoms.

nutrients: substances that nourish plants or animals or other living things

oxide: a compound containing oxygen and another element

oxygen: an odorless, colorless gas required for burning, breathing, and plant and animal life

protein: a substance made from amino acids that is part of all animals and plants and an essential nutrient

sodium acetate: a compound containing a metal called sodium and acetate, an ion which is found in vinegar

solid: something that has a definite shape and takes up a definite volume

spores: the reproductive stage some living things, such as fungi and yeasts

vitamins: a group of substances found in foods which, in small amounts, are essential to the nutrition of most plants and animals

whey: the watery material left behind when cheese is made from milk

yeasts: single-celled fungi used in baking and brewing wine, beer, etc.; they release carbon dioxide

Index

A

Acidic foods, 45, 63
Acids and bases, 17, 1819, 20, 22, 63
Air, 13, 64–65
Anti-oxidant, 63
Apples, 62–63
Artificial flavors, 71
Avocado, 54–55

B

Baking powder and baking soda, 19, 44–45
Balloons, 25–26
Bean plants, 52–53
Bird feeder, 72–73
Brown bags, 40–41
Brown sugar, 29–30
Butter, 38–39, 41

C

Cabbage juice, 18–19, 20, 22
Carbon dioxide, 26, 45, 47
Carrots, 50–51
Celery stalk, 60–61
Change of state, 35
Chemical change, 35, 37
Citric acid, 48–49
Corn plant, 56–57
Copper and copper oxide, 16–17
Cream, 38–39
Crystals, 49

D

Dormant seeds, 57

E

Eggs, 27–28

F

Fats, 30, 40–41
Food additives, 70–71
Food coloring, 58–59, 60–61
Fungi, 23, 26

G

Geotropism, 53
Glossary, 9, 78

I

Ice cream, 33–35, 74–75
Ice, 5, 15, 42–43
Indicator, 18–19, 20, 22
Infusion, 15

L

Lab in a box, 8
Lab coat, 5
Labels, 70–71
Leaveners, 45
Lemon juice, 21–22, 63

M

Margarine, 39
Materials, 7
Milk paint, 66–67
Milk, 31–32, 36–37, 38–39, 63
Molecules, 32
Mushrooms, 23–24, 26

P

Paint, 66–67
Pennies, 16–17

Physical change, 35
Pinecone, 72–73
Plants, 50–57
Play-dough, 46–47
Popcorn, 10–11, 56–57
Potato, 12–13

R

Raisin elevators, 77

S

Safety rules, 6–7
Salt, 10–11, 28, 42–43, 48–49
Salt-and-pepper trick, 77
Sea salt, 48–49
Seeds, 55, 57
Spoon trick, 76
Spores, 23–24
Sprouting, 52–53
Stamps, 68–69
Sugar, 29–30
Supplies, 7

T

Taste of food, 74–75
Tea, 14–15
Toast, 9

V

Vinegar, 17, 37

W

Water, 28, 42–43
Whey, 36–37
Yeast, 23, 25–26, 47